Autonomous Maintenance for Operators

SHOPFLOOR SERIES

Autonomous Maintenance for Operators

Edited by the Japan Institute of Plant Maintenance

Productivity Press

Additional copies of this book are available from the publisher. Discounts are available for multiple copies through the Sales Department (888-319-5852). Address all other inquiries to:

Productivity Press
444 Park Avenue South, 7th Floor
New York, NY 10016
United States of America
Telephone: 212-686-5900
Telefax: 212-686-5411
E-mail: info@productivitypress.com

Book and cover design by William Stanton
Cover illustration by Gary Ragaglia, The Vision Group
Cartoons drawn and adapted by Gordon Ekdahl, Fineline Illustration
Tables and page composition by William H. Brunson, Typography Services
Printed and bound by Sheridan Books in the United States of America

Library of Congress Cataloging-in-Publication Data

Jishu hozen no susumekata. English.
     Autonomous maintenance for operators / edited by the Japan
   Institute of Plant Maintenance ; [translated by Andrew P. Dillon].
        p. cm. — (Shopfloor series)
     Includes bibliographical references (p.).
     ISBN 1-56327-082-X
     1. Plant maintenance.   I. Nihon Puranto Mentenansu Kyōkai.
   II. Title.   III. Series.
   TS192. L57   1997
   658.2'7'0288—dc21   97-17429
                                                                       CIP

08 07   10 9

Contents

CHAPTER 6: USING ONE-POINT LESSONS 103

Publisher's Message

Total productive maintenance (TPM) is a companywide approach for eliminating equipment-related losses such as breakdowns, quality defects, and slow changeovers. Autonomous maintenance is an especially important pillar of TPM because it enlists the intelligence and skills of the people who are most familiar with factory machines: equipment operators. Through autonomous maintenance, we learn to value our machines as production partners; keeping them operating smoothly helps make our daily work trouble-free and satisfying.

The material in *Autonomous Maintenance for Operators* was originally developed for operators at Nachi-Fujikoshi Corporation, a top company in the machining industry and a winner of the PM Prize for excellence in its maintenance program. The book doesn't attempt to cover the steps of autonomous maintenance in depth, since these basics are better left to a training program specifically tailored to the needs of your company. The real value of this book is that it supports some of the standard autonomous maintenance activities through which operators can make a real difference in equipment effectiveness.

Chapter 1 of the book is about basic skills operators may learn as part of an autonomous maintenance program. The very first skill is to be able to tell when machine conditions aren't normal—an ability that continues to grow throughout the autonomous maintenance process. From this solid foundation, we can learn next what to do when abnormalities occur. We develop clear standards for equipment operating conditions and maintain those conditions through routine inspections. Along the way, we may also learn how to make simple improvements to prevent accelerated deterioration (early wear). We learn about the machine's structure and subsystems, and if it's appropriate in our situation, we may even learn how to make simple repairs.

Chapter 2 is about cleaning and inspection—the most basic ongoing activity of autonomous maintenance. Through an intensive initial cleaning, we really get to know the machine and may discover things we never realized were there. As we clean, we inspect for abnormal conditions, such as leaks or wear, and tag them for correction. When this cleaning is done, we have a much better idea of the correct conditions that need to be maintained. Then, through routine cleaning and inspection, we can spot any abnormal conditions and can take action to correct them before breakdowns or defects can occur.

Chapter 3 focuses on containment of scattering debris such as cutting chips and lubricant. This debris speeds up deterioration of cutting tools and other machine parts and contributes to quality defects. Some companies try to solve the debris problem with enclosures that surround the machine, but such covers may not keep debris out of critical functional parts; they also make regular cleaning more difficult. This chapter gives ideas for building smaller guards and covers that contain the debris locally where it is produced, leaving other functional parts clean and accessible.

Chapter 4 gives some basic ideas for a shopfloor lubrication management system. Lubrication is handled by the maintenance department in some companies, but operators know their machine conditions better than anyone else, so companies implementing TPM often involve operators in lubrication as part of autonomous maintenance. This chapter touches on basic aspects of lubrication—such as knowing how much lubricant to use—and shares ideas—such as simple color coding of inlets to match the colors of various lubricant storage containers.

Chapter 5 provides pointers for basic autonomous maintenance teamwork. Teams accomplish improvement goals through meetings and projects, supported by activity boards that remind everyone of the goals and issues and chart progress made. Chapter 6 details another aspect of team activities: learning through one-point lessons. These are brief presentations on a single sheet of paper that pass on basic knowledge about equipment systems, or describe a problem to avoid or an improvement idea that can be used more widely.

Autonomous Maintenance for Operators uses many examples drawn from the machining environment at Nachi-Fujikoshi's operations. Readers from other industries may need to adapt some ideas for their own environment, but the basic activities of autonomous maintenance can be applied in any industry to develop a smoother-running workplace that everyone is proud to be part of.

We would like to express appreciation to the Japan Institute of Plant Maintenance for permitting us to publish this English edition of their original book, with special thanks to Kunio Shirose of JIPM and the contributors from Nachi-Fujikoshi Corporation for their development of the original material.

Thanks also to Andrew P. Dillon for translation of this material; to Gordon Ekdahl for cartoon production; to Jane Loftus for copyediting; to Bill Brunson for tables and page composition; and to Gary Ragaglia for the cover illustration. Special thanks to Russ Phillips and the TPM Team of Warn Industries for reviewing the manuscript.

Many people within Productivity helped create this book. Thanks to Diane Asay, editor in chief; Karen Jones, senior development editor; Connie Dyer, TPM development director for

Productivity, Inc.; Mary Junewick, prepress editor; Bill Stanton, designer and art director; Carla Refojo, cover production; and Jessica Letteney, prepress manager.

Preface

This book brings together in a single volume six pamphlets published as the *Autonomous Maintenance Handbook Series*. From the start, this series enjoyed a wide readership and many companies urged us to publish the entire series in book form. We published this book both to respond to those desires and to reach a wider audience.

The original material from the *Autonomous Maintenance Handbook Series* was drafted by key members of the TPM promotion staff at Nachi-Fujikoshi Corporation, a company that won the PM Distinguished Plant Prize, the top international honor for maintenance program excellence. Here is how the book is organized:

Training and nurturing operators with strong equipment skills is one of the goals of autonomous maintenance. Chapter 1 describes such operators and a sequence of skill development used at Nachi-Fujikoshi. Chapter 2 presents specific and practical ways to implement the idea at the heart of Step 1 of autonomous maintenance: "cleaning is inspection." Chapter 3 introduces examples of localized containment of debris, a distinctive feature of autonomous maintenance at Nachi-Fujikoshi and a major new perspective in implementing Step 2 of autonomous maintenance (dealing with sources of contamination). Chapter 4 discusses lubrication management for the shop floor, one of the basic activities to be organized in the context of autonomous maintenance. Chapters 5 and 6 present the three classic tools of autonomous maintenance implementation: meetings, activity boards, and one-point lessons.

It is important that readers realize that the examples in this book come from the machining environment of Nachi-Fujikoshi. Therefore, some sections may have limited relevance to those who work in process industries.

The number of companies implementing TPM has been increasing for several years, and the Japan Institute of Plant Maintenance has responded by publishing reference works aimed at a broad range of organizational functions and levels. We hope this book will serve the needs of people who are implementing autonomous maintenance.

Finally, we would like to take this opportunity to express our sincere thanks to Nachi-Fujikoshi Corporation for their permission to use the material in this book.

Kunio Shirose
Executive Vice President and Assistant Director
    of TPM Operations
Japan Institute of Plant Maintenance

Contributors

Chief editor: Kunio Shirose, Executive Vice President,
Assistant Director of TPM Operations, JIPM

Ch. 1:   Yoshifumi Kimura, Engineer, TQC Promotion Office,
          Nachi–Fujikoshi

Ch. 2:   Yasuo Nosaku, TPM Promotion Deputy Manager,
          Bearing Factory, Nachi–Fujikoshi

Ch. 3:   Shoji Taniguchi, PM Manager, Machine Tool Factory,
          Nachi–Fujikoshi

Ch. 4:   Akira Tanaka, PM Assistant Manager, Machine Tool Factory,
          Nachi–Fujikoshi

Ch. 5:   Ryôei Yoshida, Manager, TQC Promotion Office,
          Nachi–Fujikoshi

Ch. 6:   Yoshihiro Mitome, Engineer, TQC Promotion Office,
          Nachi–Fujikoshi

Becoming Equipment-Conscious Operators

CHAPTER 1

CHAPTER OVERVIEW

Becoming Equipment-Conscious Operators

- Are These Problems Familiar?
- Ability to Detect Abnormalities
- Ability to Correct Abnormalities and Restore Functioning
- Ability to Set Optimal Equipment Conditions
- Ability to Maintain Optimal Equipment Conditions
- Autonomous Maintenance Step by Step
- Skill Level 1: Recognizing Deterioration and Improving Equipment to Prevent It
- Skill Level 2: Understanding Equipment Structure and Function
- Skill Level 3: Understanding the Causes of Quality Defects
- Skill Level 4: Performing Routine Repairs
- An Upward Spiral of Operator Skills
- Companies Thrive Where People Do
- Chapter Summary

Are These Problems Familiar?

In the manufacturing industry, we use machines to make products in our workplaces. Sometimes, however, the work doesn't go as planned. Do any of these situations sound familiar?

- Equipment breaks down.

- Setups and adjustments take too long.

- Running machines at the rated speeds generates defects.

- Machines are plagued by minor stoppages.

The result? Delayed shipments and missed production schedules.

Are operators in your workplace encouraged to discover problems in their production equipment? Do they have the skills needed to detect them?

Or do equipment problems go unnoticed, perhaps hidden by dirt and grime in the work area?

Ability to Detect Abnormalities

Overcoming problems like the ones listed on the previous page requires equipment-conscious operators who have four different equipment-related skills:

1. Ability to discover abnormalities

2. Ability to correct abnormalities and restore functioning

3. Ability to set optimal equipment conditions

4. Ability to maintain optimal equipment conditions

This chapter explains why these skills are important and tells how to develop them.

The ability to detect abnormalities is the first skill operators need to develop during autonomous maintenance. We have no trouble recognizing functional breakdowns, defects, and other abnormal results when they occur. To become equipment-conscious operators, however, we need to be able to see further. We need to be able to recognize abnormal conditions before the machine breaks down or produces defects.

As operators, we have the most contact with the equipment. That puts us in an ideal position to discover abnormalities before they turn into serious problems.

The ability to correct abnormalities and restore functioning

Correcting it yourself   Reporting the problem

Ability to Correct Abnormalities and Restore Functioning

Of course, merely detecting abnormalities is not enough to elim-inate equipment losses. The second essential attribute of equip-ment-conscious operators is the ability to correct abnormalities and restore functioning.

A machine can do its job only when we find the causes of "funny" or "odd" performance and then restore the original, proper state of operation. Even if we don't know how to restore correct operation, we must at least know how to explain clearly what happened to supervisors or maintenance personnel.

5

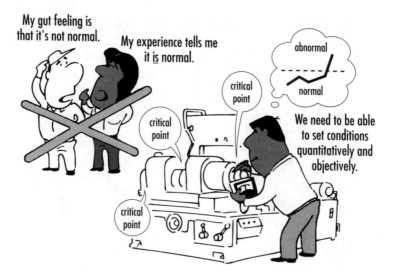

Ability to Set Optimal Equipment Conditions

Being able to tell that something is wrong depends on experience; not everyone is equally adept at spotting an abnormal situation or recognizing its importance in time to take corrective action.

For these reasons, the third requirement for an equipment-conscious operator is the ability to establish appropriate equipment-condition standards that distinguish abnormal conditions from normal ones. These standards should allow you to determine, quantitatively, whether or not the equipment's critical systems are functioning normally.

Don't put off setting standards because you're not sure what the correct conditions are. It is better to set temporary standards based on your understanding of current equipment performance, then follow these standards and revise them repeatedly until they serve as accurate condition indicators.

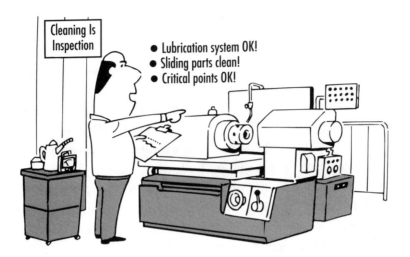

Ability to Maintain Optimal Equipment Conditions

To work with confidence, we need to be able to maintain equipment in its correct state through strict enforcement of conditions. The fourth requirement for an equipment-conscious operator is this ability to maintain optimal equipment conditions.

There are two vital elements to this ability. One is that individual operators perform daily preventive measures (making sure, for example, to add the correct amount of lubricant in the right locations and to keep critical moving parts clean). They should also monitor the equipment to make sure standards for all equipment conditions are met.

These actions allow us to work with confidence, knowing that we are getting the most out of our equipment.

Autonomous Maintenance Step by Step

Operators learn the maintenance skills they need through a seven-step program called *autonomous maintenance*. The word *autonomous* means "independent." Autonomous maintenance refers to activities designed to involve operators in maintaining their own equipment.

Steps 1, 2, and 3 of autonomous maintenance are activities to keep the state of the equipment from deteriorating. This involves reestablishing basic equipment conditions for proper operation through regular cleaning, lubrication, and tightening bolts and screws. Also involved are activities to control factors that accelerate deterioration, such as contamination by fluids, chips, and dust, and structures that make it hard to clean, inspect, or lubricate. These routine maintenance and improvement activities are ongoing and are the foundation for all the later steps of autonomous maintenance.

Steps 4 and 5 add general inspection standards that complement the cleaning and lubrication standards set up during the first three steps. In Step 4 we learn more about equipment subsystems through general inspection training. We also implement visual controls to improve equipment inspection procedures. In Step 5 we review and streamline inspection checklists based on what we learned during general inspection. Thus, we move from preventing deterioration to measuring or monitoring deterioration, and continue to make maintenance activities more efficient.

The first five steps of autonomous maintenance focus on the "hard," mechanical aspects of equipment maintenance. In Step 6 we look beyond equipment to the entire work area and production process, straightening and organizing materials and tools, standardizing, and visually managing all work activities.

Step 7 is the beginning of truly autonomous activities. This is the stage where teams carry out maintenance activities independently and where TPM really becomes business as usual.

This book highlights those autonomous maintenance activities that help us develop strong equipment-related skills.

The Seven Steps of Autonomous Maintenance

Step	Name	Activities
1	Clean and inspect	Eliminate all dirt and grime on the machine, lubricate, tighten bolts, and find and correct problems.
2	Eliminate problem sources and inaccessible areas	Correct sources of dirt and grime; prevent spattering and improve accessibility for cleaning and lubrication. Shorten the time it takes to clean and lubricate.
3	Draw up cleaning and lubricating standards	Write standards that will ensure that cleaning, lubricating, and tightening can be done efficiently. (Make a schedule for periodic tasks.)
4	Conduct general inspections	Conduct skills training with inspection manuals and use general inspections to find and correct slight abnormalities in the equipment.
5	Conduct autonomous inspections	Prepare standard checksheets for autonomous inspections. Carry out the inspections.
6	Standardize through visual workplace management	Standardize and visually manage all work processes. Examples of standards needed: • cleaning, lubrication, and inspection standards • shopfloor materials flow standards • data recording method standards • tool and die management standards
7	Implement autonomous equipment management	Develop company policies and objectives; make improvement activities part of everyday practice; keep reliable MTBF (mean time between failures) data, analyze it, and use it to improve equipment.

Skill Level 1: Recognizing Deterioration and Improving Equipment to Prevent It

The first skills operators should aim for are

- being able to recognize deteriorated equipment conditions as problems

- mastering the basics of equipment improvement

Mastery of these skills requires hands-on learning on the machine. Only by touching the machines during cleaning and inspection can we discover whether a bolt is loose or the motor is vibrating too much. This kind of learning begins in Steps 1 and 2 of an autonomous maintenance program and continues through daily routine maintenance activities. Chapter 2 will talk more specifically about cleaning and inspection activities.

By observing equipment closely, we learn how to modify equipment in simple ways to prevent the same problems from ever recurring. This happens through repeated improvements aimed at (1) eliminating the sources of contamination and other abnormalities, and (2) making it easier to inspect or service equipment. Chapter 3 gives some examples of making this type of improvement to contain contamination.

Most importantly, the activities of this first level encourage operators to treat their machines as partners and to become protective of their well-being.

Skill Level 2: Understanding Equipment Structure and Function

Understanding the structure and function of equipment is the second level of operator skill development. Specifically, this means understanding what the vital mechanisms of machines are and knowing how to clean a machine so that it performs flawlessly. The activities that build this skill are related to Steps 3 and 4 of autonomous maintenance.

To build this understanding, we draw up simple mechanical diagrams of the equipment. We should also draw the lubrication supply routes that show how lubricant reaches the areas where it is needed.

Diagrams like these help us see more clearly what we need to do on a daily basis to prevent abnormalities. They also help us carry out detailed equipment health checks.

Using these tools helps us uncover abnormalities that can cause breakdowns and recognize when to shut down equipment for repairs before breakdown can occur.

13

Skill Level 3: Understanding the Causes of Quality Defects

The third level of skill development requires that operators learn the habit of thinking logically about how product defects occur and gain the ability to deal with the root causes of quality problems. In a machining plant, this may mean developing a better understanding of the relationship between equipment precision and product quality. This particular skill relates to one type of knowledge we might learn during general inspection in Step 4 of autonomous maintenance, but it is actually an advanced activity of the type carried out at Steps 6 and 7. Some companies may train operators in this skill; in other companies precision-related improvement activities may be carried out primarily by engineering and maintenance specialists.

To support a higher understanding of the root causes of quality problems, we supplement the diagrams drawn in Skill Level 2 to show which parts of equipment influence quality. We need to understand, for example, which product quality characteristics get worse when a particular part of the equipment deteriorates. We also need to set standard values for the limits within which equipment precision needs to be maintained.

At Skill Level 3 operators monitor trends in process control limits as well as keeping up daily cleaning and inspection for causes that may be related to defects. The crucial skill to acquire at this level is the ability to detect the abnormalities that can cause product defects. Where Level 2 focused on preventing malfunctions that cause breakdowns, Level 3 goes further and looks at malfunctions that cause deterioration and loss of precision in forming the product.

Skill Level 4: Performing Routine Repairs

The ability to perform basic repairs lies at the heart of the fourth stage of operator skill enhancement. First we must understand equipment structure and function and then learn to find abnormalities that lead to breakdowns or quality failures. At Skill Level 4 we are ready to tackle restoring the equipment to its proper state. At this skill level we learn simple disassembly and repair procedures.

Companies vary in the degree to which they expect operators to carry out equipment repairs. Some companies encourage operators to develop repair skills by offering maintenance skill training and certification. At other companies, the maintenance department carries out all repairs.

The first step in this process is learning how to perform basic tasks such as

- tightening bolts correctly

- aligning pulleys and gears

- changing bearings

- stopping fluid leaks

Maintenance department people should help operators hone these repair skills.

Basic repair skills to be learned

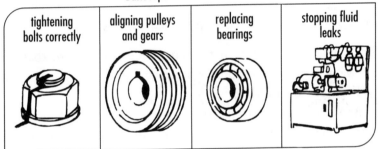

tightening bolts correctly	aligning pulleys and gears	replacing bearings	stopping fluid leaks

17

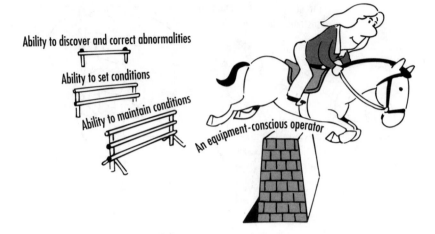

An Upward Spiral of Operator Skills

The activities described in this chapter can do more than just boost skills; they also gradually improve equipment performance and the rigor of TPM activity.

As we pay more attention to equipment, performance levels improve. This renews our confidence and interest in caring for equipment. As this leads to further improvement in the equipment, we sharpen our skills even more. These spiraling positive effects promote increasing sophistication and effectiveness in TPM activities.

An equipment-conscious
operator

Companies Thrive Where People Do

It has been observed that companies and people grow together. Growing through work is important for finding fulfillment in our jobs, and operators who develop strong equipment skills are also vital to the development of the company.

By treating machines as partners and taking responsibility for them, we get machines we can rely on and help maintain an energized and responsive workplace. Our role in reviving the original capacities of equipment is also linked to improved performance.

As you can see, equipment-conscious operators are truly the cornerstone of company growth.

19

CHAPTER SUMMARY

One of the main purposes of an autonomous maintenance program is to equip operators with skills and abilities to help prevent machine breakdowns and product quality defects. Overcoming these and other production problems requires equipment-conscious operators who have four different equipment-related skills:

1. Ability to discover abnormalities

2. Ability to correct abnormalities and restore functioning

3. Ability to set optimal equipment conditions

4. Ability to maintain optimal equipment conditions

Operators learn the maintenance skills they need to know through a seven-step autonomous maintenance program. The word *autonomous* means "independent." Autonomous maintenance refers to activities designed to involve operators in maintaining their own equipment.

The first three steps of autonomous maintenance involve activities to keep the state of the equipment from deteriorating. These include reestablishing basic equipment operating conditions through regular cleaning, lubrication, and tightening bolts and screws. Also, we take steps to control factors that accelerate deterioration, such as contamination by fluids, chips, and dust, and structures that make it hard to clean, inspect, or lubricate. These routine ongoing activities are the foundation for later steps of autonomous maintenance.

In Step 4, we learn more about equipment subsystems through general inspection training. We also implement visual controls to improve equipment inspection procedures. In Step 5, we review and streamline our inspection checklists. In these steps, we move from preventing deterioration to measuring or monitoring deterioration and continue to make maintenance activities more efficient.

The first five steps of autonomous maintenance focus on the "hard," mechanical aspects of equipment maintenance. In Step 6, we look beyond equipment to the entire work area and production process, straightening and organizing materials and tools, standardizing, and visually managing all work activities. In Step 7, teams carry out maintenance activities independently and TPM really becomes business as usual.

This book highlights autonomous maintenance activities that help us develop strong equipment-related skills. As we pay more attention to equipment, performance levels improve. This renews our confidence and interest in caring for equipment, which in turn motivates us to sharpen our skills even more.

Companies and people grow together. By treating machines as partners and taking responsibility for them, we get machines we can rely on and help maintain an energized and responsive workplace.

Cleaning Is Inspection

CHAPTER 2

CHAPTER OVERVIEW

Cleaning Is Inspection

- Initial Cleaning Is the Foundation of Autonomous Maintenance

- What Is Cleaning?

- When You Find a Problem, Tag It!

- What Are Problems?

- How Do We Look for Problems?

- Tips for Cleaning

- Tips for Inspection

- A Sample Inspection Checklist for a Hydraulic Unit

- A Sample Inspection Checklist for a Drive System

- Key Points for Removing Problem Tags

- Removing Tags: An Example

- Determining Which Tags to Remove

- Cleaning Motivates People to Take Action and to Care for Machines

- Chapter Summary

Cleaning is inspection

Initial Cleaning Is the Foundation of Autonomous Maintenance

The idea that "cleaning is inspection" lies at the heart of all autonomous maintenance activities. In a nutshell:

- Cleaning is inspection.

- Inspection is finding problems.

- Problems demand restoration to original proper conditions and improvement to prevent recurrence.

In the pages that follow, we describe specific activities that link cleaning and inspection.

25

What Is Cleaning?

Cleaning means more than simply polishing up exteriors, electrical cabinets, or covers. It means the relentless removal of years of accumulated grime on every part of a machine.

We begin cleaning by opening covers and lids, draining oil reservoirs, and touching parts of the machine we may never have seen before. This reveals many things that are not as they should be. It encourages us to think about what the machine *should* look like.

Indeed, any cleaning that doesn't expose problems and abnormalities ignores the vital link between cleaning and inspection.

When You Find a Problem, Tag It!

A thorough cleaning will reveal countless problems and conditions that are less than satisfactory. It will, in other words, be the
same as an inspection of the main body of the machine. As you
clean and inspect:

- Attach a tag to each problem you expose so that it
  won't be forgotten.

- Fix problems as soon as possible and remove their tags.

- Draw up plans for resolving problems that can't be
  addressed immediately.

This tagging activity during initial cleaning shows you how
cleaning your equipment becomes inspection as well.

27

What Are Problems?

So what problems and abnormalities do we need to check for in every nook and cranny of the machine?

There are several types of problems to look for. Some are *static* problems, like grime, scratches, looseness, and tilting, which at first glance may not seem to have anything to do with breakdowns or defects. Other problems are *dynamic,* like vibration and noise that we hear only when we run the machine.

Check for all these types of problems:

- dirt or grime
- leaks or spattering
- sagging or play
- missing or removed parts
- warping or wear
- rust or scratches
- tilting or off-center parts
- abnormal movement
- vibration or shaking
- unusual sounds or abnormal heat
- unusual smells or discoloration

How Do We Look for Problems?

We discover problems through our senses, for example by seeing, listening, smelling, and touching the machine. When it comes to finding problems, our senses are the most sensitive detection devices we have.

29

As thorough cleaning activities remove grime from the equipment, touching every nook and cranny begins to reveal problems. We become curious about other parts of the machine and open up sections of it we have never seen before. As we do, more problems appear.

Tips for Cleaning

1. Don't be afraid to get your hands dirty during cleaning.

2. Remove years of built-up grime.

3. Open covers and lids on areas you have never checked before. Clean the dirt off every hidden recess of the equipment.

4. Don't clean just the main body of the machine. Remove the grime from auxiliary equipment such as conveyors, gauges, electrical cabinets, and the insides of oil tanks.

5. Don't be discouraged by the thought that the machine will just get dirty again. Instead, focus on understanding *where* and *when* it begins to get dirty.

6. And the most important thing to remember: Clean the machine you've chosen thoroughly. Don't leave the job half done!

Tips for Inspection

1. Don't stop at problems you can see. Look for loose fastenings, slight vibrations, abnormal temperatures, and other problems that can only be detected by touch.

2. Watch for worn pulleys and belts, clogged suction filters, grime on sliding surfaces, and other problems that may lead to functional failures.

3. Are cleaning, lubricating, and checking easy to do? Do any large covers get in the way? Are oil inlets located conveniently?

4. Do all gauges work correctly, with the standard values clearly indicated?

5. Find the sources of air and oil leaks.

These are just a few examples of items you might put on a check-list for routine equipment inspections. The important point is that repeated cleaning and inspection teach us through experience to recognize problems when we see them.

During Step 4 of autonomous maintenance, operators participate in general inspections and learn more about various equipment subsystems. As a result of this training, we incorporate more specific checkpoints into system inspection checklists. The following pages show examples of such checklists from a machining plant.

31

A Sample Inspection Checklist for a Hydraulic Unit

1. Look

☐ Is the hydraulic unit dirty with chips or cutting oil?

☐ Is oil leaking from pumps, solenoid valves, joints, etc.?

☐ Are pressure gauges suitable? Do they show the right values?

☐ Are pumps and motors of the correct types and capacities?

☐ Are oil-level gauges clearly visible?

☐ Is there the right amount of hydraulic oil? Is it the right color?

☐ Are oil inlet caps on tight?

☐ Are there any gaps or holes where dirt could enter the tank?

☐ Is leaked oil returned directly to the tank?

2. Listen

☐ Are there any strange noises from pumps, motors, solenoid valves, plumbing, or other machine elements?

3. Touch

☐ Is there any unusual heat or vibration from pumps, motors, or solenoid valves?

☐ Are lock nuts coming loose on pumps, motors, or solenoid valves?

☐ Has the oil been wiped off to confirm leaks?

☐ Do hoses cross or touch one another? (Are they rubbing against one another?)

33

4. Disassemble

☐ Are there any signs of wear in motor or pump couplings?

☐ Are the vents on oil inlet caps in good condition?

☐ Are oil inlet overflows broken or dirty?

☐ Has dirt accumulated inside tanks?

☐ Are suction filters clogged?

A Sample Inspection Checklist for a Drive System

1. Look

☐ Have chips or cutting oil contaminated the motor or brake?

☐ Are belts wobbling? Are they at the proper tension?

☐ Is the brake's oil-level gauge readily visible?

☐ Are belts and pulleys shielded by safety guards that permit easy checking?

2. Listen

☐ Are there any strange noises (whining? groaning? sounds of slippage?) from the motor, brake, belts, or chains?

3. Touch while in operation

☐ Is there heat or vibration from the motor or brake?

4. Turn off the machine, then touch again

☐ Are the volume and color of oil in the brake satisfactory?

☐ Are safety guards fastened securely?

☐ Are the motor and brake mounting bolts tight?

5. Remove covers and verify

☐ Are belt tensions satisfactory?

☐ Are there the right kind and proper number of belts and pulleys?

☐ Are belts or pulleys worn?

☐ Is there any play in pulley set bolts or keys?

☐ Are belts twisted between pulleys?

☐ Are the motor and the brake properly aligned?

☐ How much wear is there on motor and brake couplings?

☐ Is there any debris blocking the motor cooling fan?

☐ Is there any debris in brake lubricant?

Key Points for Removing Problem Tags

1. Whenever possible, the people who discover and tag equipment problems should be the people who fix them.

2. Approach every problem with a vision of how the machine *ought to be*.

3. Don't overlook even the slightest defect. Fix it out of principle, without worrying about what effect it might or might not have.

4. Read the operating manual at least five times to make sure you understand it thoroughly.

To remove problem tags, we need to get our hands dirty and use our ingenuity and concentration to make simple repairs we may never have made before. In doing this, we learn where contamination occurs and what the machine should look like, and think again about the natural tendency to ignore small problems. Struggling to fix problems inspires us to find ways to prevent them from ever occurring again.

Removing Tags: An Example

Tag text:

> Oil leaks from hydraulic lathe feed-cylinder rod every time the rod moves.

Removing the tag means more than simply fixing the problem. We need to trace the problem back to its root cause by asking "why" repeatedly:

No.	Items to Check	Tags Issued	Tags Removed
1	Oil pressure?	Pressure is too high	Pressure corrected
2	Cylinder movement? →	Too slow, doesn't move	
3	Reason for movement problem? →	Clogged strainer	Cleaned strainer
4	What clogged strainer? →	Oil was dirty	Removed oil and cleaned
5	What contaminated the oil? →	Dirt got into tank	Prevented chip and oil spatter
6	How did dirt get in? →	There was a hole or gap in tank cover	Filled in all holes and gaps
7	How about movement? →	OK, but oil is leaking	Disassembled leaking sections
8	Why did it leak? →	O-ring is broken	Changed O-ring
9	Why did it break? →	Rod is nicked	Removed nick
10	Why was it nicked? →	Chips flew in and stuck to rod Dirt in oil caused problem	Took steps to prevent spatter of chips Went back to (5)
11	Movement? Leaking? Oil temperature?	OK None OK	Wrote up one-point lesson

This series of attachment and removal steps cannot be done all at once. Only repeated attention to the problem will change the way we see it and improve the actions we take.

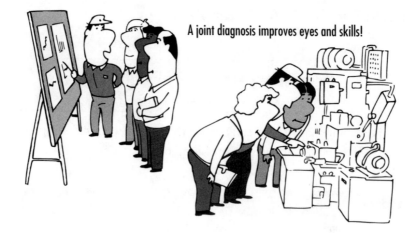

A joint diagnosis improves eyes and skills!

Determining Which Tags to Remove

How can we best determine which tags we can remove ourselves and which tags require outside assistance?

In one leading factory, an autonomous maintenance diagnosis team made up of managers and maintenance and engineering specialists goes to the machine and, together with the operators, discusses methods used for cleaning and inspection and for tag attachment and removal. This joint approach supports the development of operator skills by focusing attention on problems and by offering examples and guidance.

Participating in a joint diagnosis at each stage of autonomous maintenance will improve the *seeing* and *doing* skills of operators as well as diagnosis team members.

Cleaning Motivates People to Take Action and to Care for Machines

In spite of the effort that we put into cleaning, machines will get dirty again. The more thoroughly we clean, the more we will want to find ways to keep equipment from getting dirty again. This is the kind of motivation that results in, for example, antis-catter guards that engineers would never have thought of. (See Chapter 3 for more containment ideas.)

Once relentless cleaning and inspection have exposed the effects of accelerated deterioration (described in Chapter 3), we can restore equipment to its original condition and maintain basic standards for cleanliness, lubrication, tightness, and other critical operating conditions. The result on the shop floor is an outpouring of confidence and determination.

39

CHAPTER SUMMARY

Initial cleaning of equipment is the heart of autonomous maintenance. As we clean, we inspect the equipment for abnormalities that could cause bigger problems in the future. When we notice abnormalities, we take action to fix them and make improvements so that they don't happen again.

Cleaning in autonomous maintenance means deep cleaning and inspection of every part of the machine. This inspection will naturally turn up abnormalities and problems. To make sure they are not forgotten, we attach a tag to each problem we expose. If we can fix the problem easily, we may do it ourselves, or we may need to ask for maintenance assistance.

During inspection, we look not only for static problems that are visible even when the machine is not running, but also for dynamic problems that appear during operation. Important types of problems to check for include

- dirt or grime
- leaks or spattering
- sagging or play
- missing or removed parts
- warping or wear
- rust or scratches
- tilting or off-center parts
- abnormal movement
- vibration or shaking
- unusual sounds or abnormal heat
- unusual smells or discoloration

Cleaning and inspection mean using our senses, including seeing, listening, smelling, and touching the machine. It is a hands-on process that should reach every area of the machine, using simple checklists to cover various parts and systems. During general inspection in Step 4 of autonomous maintenance we will learn more about specific equipment subsystems and make more detailed inspection checklists.

Whenever possible, the people who discover and tag equipment problems should be the ones to fix them and remove the tags. Sometimes it is helpful to have a diagnostic team that includes managers and maintenance and engineering specialists review the tags to help determine what to do about them. This joint approach helps operators develop equipment skills by focusing attention on problems and by offering examples and guidance.

The more thoroughly we clean, the more we will want to find ways to keep equipment from getting dirty again. This kind of motivation results in unique solutions to eliminate sources of debris, such as operator-built anti-scatter guards that engineers might never have thought of. (Chapter 3 shares examples of such containment ideas.)

Using Localized Containment

CHAPTER 3

CHAPTER OVERVIEW

Using Localized Containment

- What Is Accelerated Deterioration?

- The Trouble with Large Enclosures

- The Point of Containment

- Tips for Containment

- Continually Improving Containment

- Types of Containment Guards

- Example: A Containment Guard for a
  Centerless Grinder

- Example: Containment Guards for a Cylindrical
  Grinder and a Superfinisher

- Example: A Containment Guard for a
  Worm Grinder

- Example: A Containment Guard for an NC Milling
  Machine

- Example: Methods for Controlling Oil Flow

- Decreasing Breakdowns and Defects with Localized
  Containment

- Chapter Summary

Accelerated deterioration happens
when people don't do
what they need to do.

What Is Accelerated Deterioration?

Deterioration is what leads equipment to break down or to generate defective parts. There are two kinds of deterioration: *natural deterioration* and *accelerated deterioration.*

Even when equipment is used correctly, certain parts rub against one another, gradually causing physical wear. We refer to this as *natural deterioration.*

By contrast, *accelerated deterioration* is deterioration that happens sooner than it would naturally. Accelerated deterioration is usually caused by our failure to do something we ought to do—for example, by not keeping parts clean and lubricated or by ignoring excessive loads or play in moving parts.

45

The Trouble with Large Enclosures

Machines on our shop floors spew out dirt and grime in the form of chips, oil, and stripped metal. Wet-type processing equipment, in particular, throws off enormous quantities of chips and coolant, and a machine that has just been cleaned rapidly gets dirty again. This debris is a leading cause of accelerated deterioration.

Some workplaces try to solve the problems caused by scattered debris by using large enclosures or guards that, in some cases, cover the entire machine. Such enclosures, however, suffer from several drawbacks:

1. They do nothing to prevent actual accelerated deterioration and can be linked directly to breakdowns.

2. They make maintenance harder by making it difficult to clean, lubricate, and check equipment.

3. They complicate changeovers and lengthen changeover times.

The Point of Containment

In addition to the shortcomings already mentioned, enclosure-type guards are costly to build, take a long time to clean, and can bring subsequent improvements to a halt.

How can we overcome these drawbacks? The key lies in two ideas: making guards as small as possible (localization) and bringing them as close as possible to the source of contamination to absorb the scattered material (containment). This is one of the strategies we use to implement Step 2 of autonomous maintenance, which focuses on removing the sources of problems and eliminating inaccessible areas.

When we fail to install guards or covers to stop dirt and grime from scattering into the work area, we cause breakdowns by accelerating the deterioration of critical parts of the equipment. Use your ingenuity to devise guards and covers that will limit equipment deterioration to the "natural" kind.

Using localized containment to control scatter produces both tangible and intangible benefits:

1. Accelerated deterioration stops because dirt and grime don't penetrate to the critical functional parts of equipment. This reduces breakdowns considerably.

2. Cleaning can be performed quickly.

3. Easy lubrication and checking promote thorough maintenance.

4. The application of cutting oil and other machining conditions can be checked easily.

5. Setups and changeovers go quicker because the machine is cleaner and not obstructed by large guards.

6. Team members are energized and motivated to take up the challenge of difficult improvements.

7. Leaders and operators alike learn the techniques—and experience the benefits—of continuous improvement.

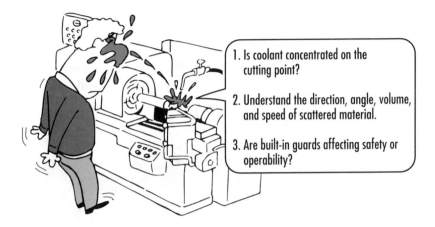

Tips for Containment

Here are some tips for containing chip and coolant scatter on wet-type machining equipment.

1. First of all, check whether the coolant is concentrated on the cutting point. When it is not, scatter gets much worse, tool life shortens, and the quality of the workpiece finish (for example, surface roughness) deteriorates. This situation occurs surprisingly often, so we need to observe and confirm what is really happening.

2. Understand the direction, angle, volume, and speed of the scattered material.

3. Check the shape and size of built-in guards and notice how they may affect machine safety and operability.

49

Continually Improving Containment

The operator of a machine knows about the scatter of chips and coolant better than anyone else. This is why operators themselves should build and install temporary "homemade" covers and guards made from corrugated cardboard or sheet metal.

The solution doesn't need to be perfect from the start. And you don't need to bring in the engineers to help you. Building temporary localized guards is one way for team members to practice relying on their own ideas and skills.

The best process is one of study and improvement. The team uses its homemade guards, observes their effectiveness, adjusts the design to correct any problems, and then builds more durable versions.

a shield-type guard – inexpensive and quick to make

After testing, make a more permanent version from plastic, rubber sheeting, or sheet metal.

a box-type guard – costs more, but has bigger impact

Types of Containment Guards

Containment guards can be divided into two categories based on their shapes. Both types should stop scatter close to the source of the problem. The first type is a simple shield that consists of one or more plates positioned next to the critical site. These don't cost much and can be built quickly.

Then there are box-type guards, which are compact or localized versions of large antiscatter covers. These are more expensive, but they can be surprisingly effective to keep debris in or out. There are many possible designs; the essential elements depend on the structure of the mechanisms and the creativity and persistence of shopfloor operators.

Each guard or cover should be tested in cardboard or sheet metal before a more permanent version is built out of plastic, rubber sheeting, steel, or other material.

51

Shield-type containment guards made of PVC plastic, rubber sheeting, and sheet metal (mist still comes out from gaps)

PVC box-type containment guard (no more mist)

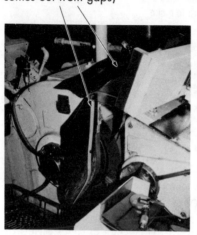

Photo 1

Photo 2

Example: A Containment Guard for a Centerless Grinder

52

The centerless grinder in this example was equipped with an automatic feed mechanism. The small gap between the grinding wheel and the feed grindstone meant that a lot of grinding lubricant needed to be applied, making it difficult to contain scatter. At first the team installed simple shield-type containment guards (see Photo 1).

Gaps in these shield-type guards still allowed a lot of mist to escape, however, so the team members observed the situation carefully. They found that the flat, open-mouthed nozzle for the grinding lubricant missed the grinding point and caused a lot of scattering.

The team finally shut out scatter by substituting several pipe-shaped nozzles that aimed the lubricant more precisely. They also switched to a box-type containment cover (see Photo 2).

Photo 3

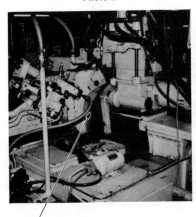

Photo 4

A vinyl guard next to the source of the debris controls scatter.

A plastic box over the honing head contains scatter.

Example: Containment Guards for a Cylindrical Grinder and a Superfinisher

Most cylindrical grinders are equipped with a cover to prevent grinding lubricant from scattering off the grindstone. These covers generally make it more difficult to clean, lubricate, and check the machine. They also add time and trouble to grinding wheel changeovers.

Simply installing vinyl shield-type guards on the grinding wheel side near the source of contamination made it possible to remove the antiscatter cover (see Photo 3).

In the same way, installing box-type guards on superfinishers with 3-axis honing heads changed a dirty, slick bed to one that was dry and not slippery (see Photo 4). This also solved the cleaning problems that go along with a large cover.

53

Example: A Containment Guard for a Worm Grinder

Now let's look at an example where a team changed its thinking and switched from a large antiscatter enclosure to a compact box-type cover.

The large cover that originally surrounded the machine (see Photo 5) would fill up with mist and grinding dust, which got into belts, couplings, and limit switches, as well as the equipment's sliding surfaces. This accelerated the deterioration of all these parts, to the point where some 30 machines broke down about five times each month.

Operators also struggled with these machines, because they had to open the covers every time they needed to load or unload parts or to check on the grinding process.

Even worse, the machines would get dirty again immediately after cleaning. The cleaning, lubricating, and checking standards established during the early steps of autonomous maintenance were nearly impossible to maintain.

These pressing needs spurred the improvement team to develop smaller, box-type containment covers for specific scatter sites. The critical functional part at the top of the machine was kept open for easier cleaning and maintenance. This unprecedented improvement resulted in both tangible and intangible advantages (see Photo 6).

The whole machine was enclosed in a big cover, so the inside filled up with chips (accelerated deterioration was neglected).

Photo 5

The critical section is kept open and accessible.

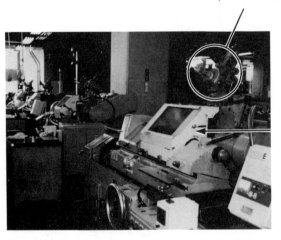

Photo 6

Here, a box-type containment guard encloses just the section where chips are generated. This keeps chips from entering the crucial functional part of the equipment (accelerated deterioration has been reduced to natural deterioration).

Example: A Containment Guard for an NC Milling Machine

In this example, operators developed containment guards to prevent scattering and contamination by chips on a numerically controlled milling machine.

The variety of workpiece shapes, dimensions, and materials had frustrated attempts to prevent scattering and contamination; as a result, chips piled up around the machine and underfoot as well as inside the machine. Cleaning up the mess took 35 minutes each day, and chips that got into the lubricant tank caused as many as three breakdowns each year.

The team approached this situation by installing rubber curtain-type guards around the cutting tool where the chips originated. These guards kept chips from flying long distances, but chips bouncing off the guards still piled up, so the team next installed a table cover (see Photo 7).

But even this did not stop chips entirely. The team finally put a stop to accelerated deterioration by installing a retracting-type cover around the critical table surfaces.

Photo 7

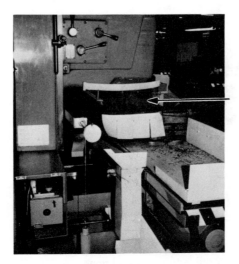

Chip scatter is prevented
by the installation of an
open/shut containment
guard.

open/shut containment guard
made of rubber, PVC,
and sheet metal

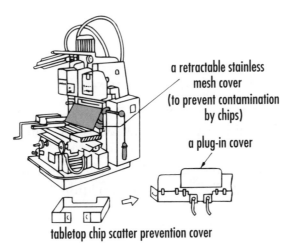

a retractable stainless
mesh cover
(to prevent contamination
by chips)

a plug-in cover

tabletop chip scatter prevention cover

Photo 8

an L-shaped guide to control oil flow

Example: Methods for Controlling Oil Flow

Improvements to control oil flows in wet–type machining equipment are similar to the containment approach. Rather than let-
ting used grinding oil fill the machine's oil pan, one team used
silicone sealant to make channels that direct oil and chips to a
limited area.

This kills two birds with one stone:

- It shortens the time it takes to clean chips and grind-
  ings in the machine as a whole.

- It also improves the appearance of the machine and
  speeds up the recycling of oil, so less grit builds up.

Photo 8 shows another use of this material: an L-shaped guide
plate caulked with silicone to prevent oil from leaking through.

Decreasing Breakdowns and Defects with Localized Containment

As localized containment is installed to control scatter, natural deterioration patterns replace accelerated deterioration and the number of machine breakdowns usually drops dramatically.

When breakdowns become rarer and take up less time, quality defects in the form of scrap and rework typically fall as well. When you take on the challenge of scatter containment, you not only cut breakdowns and defects, but gain gleaming machines and a bright workplace.

59

CHAPTER SUMMARY

Preventing accelerated deterioration is a key purpose of autonomous maintenance. Deterioration refers to wear that eventually causes equipment to break down or produce defects. Some deterioration is natural, even when equipment is used correctly. Accelerated deterioration is deterioration that happens sooner than it naturally would. It usually results from failure to do something we ought to do—for example, not keeping parts clean and lubricated.

Keeping the work area clear of scattered debris such as cutting chips is an important concept in reducing accelerated deterioration. Some plants try to contain debris by using large enclosures or guards that may cover the entire machine. However, this kind of guard doesn't actually prevent accelerated deterioration. In fact, it makes the equipment harder to clean and inspect, as well as complicating changeovers.

Localized containment is a more effective approach to controlling scattered debris. It uses two ideas: making guards as small as possible (localization) and bringing them as close as possible to the source of contamination (containment). Localized containment is a key strategy in Step 2 of autonomous maintenance, which focuses on removing the sources of problems and eliminating inaccessible areas.

Localized containment stops accelerated deterioration because dirt and grime don't penetrate to the critical functional parts of equipment. This reduces breakdowns considerably. It makes cleaning, lubrication, and inspection quicker and more thorough. Oil application and other machining conditions can be checked easily. Setups and changeovers go quicker.

In planning localized containment for machining equipment, make sure the coolant is concentrated at the cutting point. Study the direction, angle, volume, and speed of the scattered debris. Check the shape and size of built-in guards and look for any safety or operability problems.

Operators understand debris scatter problems better than anyone else. We can use this knowledge to devise temporary covers and guards out of inexpensive materials. After testing and refinement, we can make more durable versions.

There are two types of localized containment guards. One type is a simple shield that consists of one or more plates positioned next to the critical site. These don't cost much and can be built quickly. The other type, box-type guards, are compact or localized versions of large antiscatter enclosures. These are more expensive, but they can be surprisingly effective. Consider the examples in the chapter for specific ideas.

As localized containment is installed to control scatter, natural deterioration patterns replace accelerated deterioration, and the number of machine breakdowns usually drops dramatically. When breakdowns become rarer and take up less time, quality defects in the form of scrap and rework typically fall as well.

Lubrication Management
for the Shop Floor

CHAPTER 4

CHAPTER OVERVIEW

Lubrication Management for the Shop Floor

- Why Lubricate?
- The Purposes of Lubrication
- What Is Lubrication Management?
- Types of Lubricant and Their Uses
- Grouping Lubricants
- Adding Lubricant at the Right Time
- Adding the Right Amount of Lubricant
- Using the Right Method
- Operator Responsibilities and Lubrication Tips
- Daily Lubrication Checks
- Making a Lubrication Diagram
- Setting Lubrication Standards
- Toward Automatic Lubrication
- Visual Management of Lubricants
- Controlling Contamination
- Toward Zero Breakdowns Due to Lubrication Failures
- Chapter Summary

Why Lubricate?

According to some experts, 60 percent of breakdowns that occur in the moving parts of machine tools happen due to poor lubrication or oiling. There are several possible reasons for lubrication problems, including

- failure to adhere to correct lubrication standards

- absence of lubrication standards

- use of the wrong standards

65

Getting rid of lubrication problems means, first of all, never, ever running equipment without lubrication. In the first three steps of autonomous maintenance, we learn to inspect lubrication sites to make sure levels are appropriate; we also learn to recognize abnormal conditions such as excess grease and oil drips. Later, during general inspection training in Step 4, we may learn about lubrication management. We may also learn proper lubrication practices: how to add just the right amount of the right lubricant at the right time. This chapter focuses on some of the lubrication management issues that might be encountered during Step 4.

Caring about our equipment and knowing its structure and function make it possible to lubricate correctly and to wipe out problems caused by inadequate lubrication.

The Purposes of Lubrication

Physical objects moving in contact with one another causes friction resistance. We use lubricants to minimize this resistance. Lubrication also

- prevents friction and reduces wear

- eliminates heat caused by friction

- prevents rusting caused by moisture

- washes away debris such as grinding chips

Maintaining proper lubricant levels assures smooth running and extends machine life. Ultimately, it leads to increased productivity.

The **right** lubricant, in the **right** amount, by the **right** method, at the **right** time!

What Is Lubrication Management?

There is nothing difficult about lubrication management. It simply means lubricating equipment correctly and reliably. Machines need to be lubricated

- according to their use

- with clean lubricant

- at designated inlets or fittings

- at specified times

- with specified amounts of lubricant

- to ensure coverage of surfaces in contact

Lubrication management on the shop floor simply means specifying and adhering to these conditions.

Common Types of Oil and Their Uses

Types	Uses
Spindle oil	A low-viscosity oil used for high-speed, low-load rotating parts. Mostly used in small electric motors and spinning machines, etc.
Cutting oil and fluid	Used in various machining operations to lubricate and cool. Oils and fluids may contain various additives depending on the materials cut and the processing speed.
Turbine oil	High-speed bearing lubricant for all turbines; also widely used as hydraulic oil.
Machine oil	Widely used to lubricate bearings and rotating friction parts of many machines. Quality varies especially according to type of base oil.
Cylinder oil	A typical high-viscosity oil for high-load, high-temperature parts. Used principally for steam engine cylinders and valves.
Gear oil	Used in all types of gears to reduce friction and heat at low speeds. There are several types, which can be chosen according to load resistance.
Hydraulic oil	Formed of additives and a turbine oil base, this generally low-viscosity oil is used in hydraulic systems.

Types of Lubricant and Their Uses

Various types of lubricants may be used for specific purposes. Some companies may select lubricants with additives for particular hard-to-machine applications. The table shown above describes typical uses for several common categories of oil and lubricating fluid.

There are also multipurpose lubricants that can be used several ways. For example, a lubricating oil may be used as a cutting fluid on the same machine. Another example is a lubricant that can function as a hydraulic fluid. Machine manufacturers often recommend specific lubricants for optimum performance.

Grouping Lubricants

Using exactly the lubricants specified by each equipment manufacturer would require us to stock many different brands of lubricant and increase the chance of grabbing the wrong type of lubricant off the shelf. For this reason, we often group lubricants and keep a smaller number on hand.

To group lubricants:

1. Make a list of lubricant brands and types currently in use.

2. Group lubricants that have similar viscosity.

3. For each group, reduce the total number of lubricants kept on hand by using multipurpose lubricants and lubricants that will work for more than one machine. (Check with machine manufacturers to be sure the selected lubricants are appropriate.) Then decide how to store the lubricants and set up lubrication stations.

A lubrication station should hold about three containers, each of which is filled with lubricant of a specified viscosity. These and other lubrication containers and tools should be color-coded to help prevent mixing lubricants.

Adding Lubricant at the Right Time

Adding lubricant at the right time is crucial. To assure timeliness, a lubrication schedule should be established. Usually the equipment operating manual will specify when to lubricate. If the manual gives no guidance, you must set sensible standards and then verify lubrication coverage through trial and error.

To avoid running out of lubricant, set lubrication intervals slightly shorter than you think they need to be, then gradually lengthen them if necessary.

Lubricating at the right time obviously means setting a schedule and sticking to it. Generally you should be prepared to lubricate and check your equipment every day at the beginning of work.

Adding the Right Amount of Lubricant

Using too much lubricant or too little lubricant can cause lubrication defects.

Using too much lubricant can bring on defects by causing leaks, by lowering viscosity, and (depending on the lubrication method) by promoting deterioration of the lubricant.

Using too little lubricant can reduce the amount of lubrication film between the moving parts and cause scuffing and burning.

Factors in determining the suitable quantity of lubricant include

- operating conditions, such as the machine speed and the type of load

- whether the friction surfaces are sliding or rotating parts (rotating parts tend to require more lubricant)

- the lubrication method used

71

Fluid Level Standards by Lubrication Method

Lubrication Method			Fluid Level	
			Minimum Level	Maximum Level
Drip method			1/3 of height of oil reservoir	Full height of oil reservoir
Wick method			1/3 of height of oil reservoir	Full height of oil reservoir
Chain oiler			Bottom of chain should be in oil	Maximum height at which no oil will spill from reservoir during operation
Collar oiler			1/2 height of collar from bottom of axle	Bottom axle
Oil-mist method			Follow precise level indicated on oil feeder	
Mechanical force-feed method			1/2 height of reservoir or oil level gauge	Full height of reservoir or "full" on oil level gauge
Oil-bath method	Ball and roller bearings	bm n <8,000	No more than 1/10 of ring diameter above bottom of ring	
		bm n >8,000	1/2 to 4/5 of roller diameter above bottom of rollers	
Pressure-circulating method			1/2 of tank height when running	9/10 of tank height when stopped

Using the Right Method

Using the right method assures that the right amount of lubricant is applied to friction surfaces at the right time.

Using the right method means using clean lubricant, specified tools, and specified inlets or grease fittings. Check whether the fluid level is within the standard range for the type of lubrication method; the table above shows fluid level standards for several common methods. As you check, look especially for

- lubricant contamination
- contamination of lubrication tools and reservoirs
- damage or other problems with inlets

Operator Responsibilities and Lubrication Tips

Responsibilities of the Operator in Lubrication Management

1. Always keep the machine's lubrication and friction surfaces clean.

2. Keep lubrication containers, tools, and implements well organized and clean.

3. Keep container caps and lids securely closed.

4. Make sure no debris or moisture gets into lubricants.

5. Label lubrication sites with the appropriate lubricant type and frequency.

73

Tips for Using Lubricants

1. Don't allow contamination by dust, debris, or moisture, etc.

2. Don't mix lubricant types.

3. Pay attention to safety.

Daily Lubrication Checks

The most important function of daily checks is to tell us about normal operating conditions such as fluid levels, lubricant temperature, noise, and vibration.

74

1. Is lubricant getting to the parts that need lubrication? (Check actual parts periodically.)

2. Are lubricant quantities and temperature, etc., within specified limits?

3. Are there any lubricant leaks?

4. Are there any abnormalities in the quality of the lubricant (discoloration, clouding, foaming)?

5. Are there any unusual noises or vibrations?

6. Is the lubricant consumption rate higher or lower than normal?

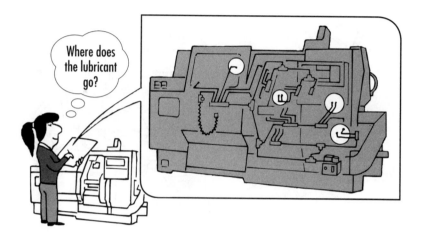

Making a Lubrication Diagram

Even if you are adding lubricant regularly, you need to check that the lubricant is really reaching the surfaces it is supposed to reach. Operators can do this themselves by diagramming or sketching the route lubricant takes between inlets and destination surfaces. This allows them to check along the route whether, for example, pipes are broken, oil is flowing correctly from distributors, joints are leaking, or oil is spilling or leaking somewhere.

Personally checking that lubricant is reaching its destination allows operators to run their machines with confidence and enhances their awareness of the importance of lubrication.

75

Setting Lubrication Standards

Lubrication standards need to express what actions *we* need to take to manage and maintain *our* machines. It is crucial that equipment operators experience the things we have been discussing and that they help determine the lubrication standards that need to be followed.

Lubrication standards need to include

1. lubricant inlet locations
2. method for checking
3. lubrication method
4. lubricant type

5. lubrication tools
6. time needed for lubrication
7. lubrication schedule
8. person responsible, etc.

When establishing lubrication standards, avoid standards and methods that are merely theoretical or ideal. Your lubrication standards must be reliable and realistic, based on actual experience and on the operating manual.

Toward Automatic Lubrication

Lubrication assures efficient equipment operation by providing lubricant at the right times and in the right quantities. But some pieces of equipment may have more than 20 places where lubricant needs to be added, overloading the operator with lubricating and checking responsibilities. When this is the case, carefully crafted lubricating and checking standards are irrelevant because there is no way to make sure that what *should* be done actually *is* done.

Such situations call for lightening the operator's burden to make sure needed lubrication is carried out. One solution is a centralized lubrication system, in which lubricant is added manually and distributed to various points on the machine. Another solution is an automatic lubrication system, in which lubricant is added automatically from a reservoir and driven to the points where it is needed.

77

A sight gauge for
oil level

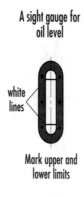

white
lines

Mark upper and
lower limits

A sample color code for lubricants and fluids

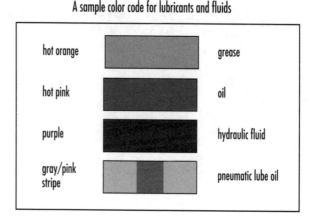

hot orange		grease
hot pink		oil
purple		hydraulic fluid
gray/pink stripe		pneumatic lube oil

Visual Management of Lubricants

Easy-to-follow visual management tools should be used for lubrication inlets, storage containers, and lubrication tools to ensure proper lubrication practices. Marking minimum and maximum levels on oil gauges is a simple example. Color coding is another important visual technique. For example:

1. Distinguish lubricants of different viscosities by color-coding their storage containers.

2. Use these same colors on the lubricant inlets on machines to indicate the type of lubricant to use. (Some companies use a second color on the inlet to indicate the frequency; other companies color code by type and follow a checklist for the frequency.)

3. Color-code your lubrication tools according to type and viscosity.

Color-coding helps mistake-proof the process, giving people an easy way to check whether they are using the right lubricant and the right tools. (Be sure to follow any color code standards established by your company or by regulatory agencies. Also write the lubricant names on containers and inlets for people who may be color-blind.)

Controlling Contamination

Neglecting dirt, grime, dust, discoloration, or deterioration in lubricants can lead to many undesirable results. Among other things, ignoring these problems can promote wear of machine parts, slow down equipment, and decrease part life. These effects are often overlooked on the production floor because they don't show up right away.

Operators need to check for contamination with their senses (especially the sense of sight). If the lubricant appears to be contaminated, clean the machine's lubricant tanks and filters, as well as the friction surfaces. Also clean the implements and tools used during lubrication.

Such activities should focus attention on specific past contamination problems and on changes and improvements to prevent contamination in the future.

79

Toward Zero Breakdowns Due to Lubrication Failures

> **Key questions for eliminating breakdowns**
>
> 1. Are we using the correct lubricant for the machine?
>
> 2. Are we using the right amount?
>
> 3. Are we ignoring leaks or overfilling?
>
> 4. Are we storing lubricants correctly?

Breakdowns due to lubrication failures can be eliminated entirely when we reexamine the workplace from the perspective of these questions. Equipment operators can play a leading role in carrying out proper lubrication maintenance by helping develop appropriate lubrication standards and then following them on a regular basis.

CHAPTER SUMMARY

Poor lubrication causes about 60 percent of all breakdowns. In the first three steps of autonomous maintenance, we learn to inspect lubrication sites to make sure levels are appropriate; we also learn to recognize abnormal conditions such as excess grease and oil drips. During general inspection training in Step 4, we may learn about lubrication management and how to add just the right amount of the right lubricant at the right time.

Lubricants minimize the friction resistance caused when physical objects move in contact with one another. Lubrication reduces wear and eliminates heat caused by friction; it may also be used to prevent rusting caused by moisture and to wash away debris such as grinding chips.

Various types of lubricants are used for different purposes. To reduce the number of brands stocked, your company may consolidate and use one lubricant for a range of viscosity requirements. Lubrication stations keep quantities of these lubricants close to the machines in which they are used. Color-code containers and lubrication tools to prevent mixing.

Adding lubricant at the right time is crucial. Establish a lubrication schedule in accordance with the equipment operating manual. If the manual gives no guidance, set sensible standards and then verify lubrication coverage through trial and error.

Using the right amount of lubricant is also important. Too much lubricant can cause leaks, lower viscosity, and lead to deterioration of the lubricant. Too little lubricant can reduce the friction-reducing film of oil and cause scuffing and burning.

81

Using the right method to add lubricant assures that the right amount is applied at the right time. Use clean lubricant and specified tools and inlets, and check the fluid level in accordance with the type of lubrication system.

Be sure to include lubrication items in the daily check of equipment conditions. Make sure lubricant is getting to the parts that need it, and that quantities and temperature are within specified limits. Look for leaks and spikes in consumption, abnormal appearance, and unusual noises or vibrations. Make a lubrication diagram to check that lubricant is reaching the destination surfaces.

When you understand the lubrication process that needs to take place, establish lubrication standards. If lubrication tasks are so complex and time consuming that they are neglected, a centralized or automatic lubrication system should be considered to make the tasks less burdensome.

Visual management can help avoid lubrication errors. Color code the lubricant containers, inlets on the equipment, and lubrication tools to ensure that the right type of lubricant is used.

Neglecting dirt, grime, dust, discoloration, or deterioration in lubricants can promote wear of machine parts, slow down equipment, and decrease part life. Control contamination by cleaning the machine's lubricant tanks and filters, as well as the friction surfaces. Keep lubrication tools clean.

Equipment operators can play a leading role in carrying out proper lubrication maintenance by helping develop appropriate lubrication standards and then following them on a regular basis.

Activity Boards and Meetings

CHAPTER 5

CHAPTER OVERVIEW

Activity Boards and Meetings

- Autonomous Maintenance and Team Activities
- Where Does Motivation Come From?
- The Activity Board: A Guide to Action
- Translating the Vision
- Displaying the Activity Board in the Workplace
- Giving Encouragement to the Team
- Remembering the Goal
- Components of an Activity Board
- Using the Activity Board in Meetings
- Managing Meetings
- Keeping Minutes of Meetings
- Rules for Meetings
- Ten Tips for Leaders
- Conclusion
- Chapter Summary

Autonomous Maintenance and Team Activities

The goal of autonomous maintenance is to develop operators with truly strong machine-related skills. Teams full of highly skilled people will build a strong, vital workplace.

Team success depends on motivation, ability, and a favorable environment. Ability comes from practice and confidence grows out of success. The team as a whole grows when all its members are motivated and encouraged to attack a problem, encourage one another, and work through to a solution. Teamwork energizes the workplace and everyone shares in the satisfaction of achievement.

85

Where Does Motivation Come From?

We are most likely to attack problems willingly when we have a clear understanding of the basic elements of problem solving:

1. *What* are we going to do? (Theme)

2. *Why* are we going to do it? (Vision)

3. *How far* are we going to go? (Targets)

4. *How* are we going to do it? (Method)

5. *What is the sequence* and timing of actions? (Schedule)

6. *Who* does what? (Roles)

7. *What results* do we expect? (Assessment)

Motivation ultimately depends on how well every team member grasps these issues.

The Activity Board: A Guide to Action

The team needs a visual guide to its activities that makes the items we've just listed so clear that anyone can immediately understand them. An activity board fills these functions and that is why it is such an effective tool.

Let's think of an analogy from baseball. Here's the situation. The two teams are tied. It's the bottom of the ninth inning, the bases are loaded, and there are two strikes and three balls. We're watching the pitcher's final movements. The batter and everyone in the dugout are probably holding their breaths. No explanation is needed. There are rules and a clear score is displayed as the game proceeds.

The same clarity is important in our work. Why not use the same approach? We will know when we have succeeded. A team activity board is our scoreboard.

87

Translating the Vision

Success in work is a source of personal satisfaction. We feel even greater satisfaction, though, when our accomplishments do something of value for others in the workplace and for the company as a whole.

This is where an activity board comes in. It tells at a glance what the vision of the company is and translates that vision into divisional and departmental goals that point to what the team can do. From there, it documents what problems need resolving, what approaches we are using, and how we have planned our actions.

Displaying the Activity Board in the Workplace

Even when a team has decided what it needs to do and has kicked off activities with a shared understanding of the problem, after a few months team members tend to forget the purpose of the project and goals drift out of focus.

A team's ability to negotiate this phase will depend on how clear and well organized its activities have been. One way to avoid losing focus is to display a large team-activity board at the group's meeting place and to hold daily morning meetings in front of the board.

To be effective, information on an activity board needs to be readable from a distance. The board should be colorful, with daily and monthly progress shown in easy-to-understand graphs.

Giving Encouragement to the Team

As a gauge of team activities, an activity board should instantly tell anyone who looks at it where and at what speed the team is heading and where current efforts are focused.

One of the main functions of an activity board is to express support for people and their persistent hard work. A board that everyone can understand will spur team members to keep up their efforts. Sometimes people need extra encouragement, and often a manager's simple "Good work!" or "Way to go!" posted on the board can give the team what it needs. This sort of encouraging message from managers often inspires teams to take on new challenges.

Remembering the Goal

When team members have done what they planned to do or achieved their improvement aims, they should not forget to evaluate the results.

When team members and leaders meet to establish the next improvement plan they should take into account a number of issues: Were the goals achieved consistent with team objectives? Did the activity resolve everything or do problems remain? Was anything left undone? How much, and why? Did the team run into problems during their activities? What is the next step?

The activity board helps the team understand these issues. It should express at a glance

- the team's goals

- specific problems the team is addressing

- actions the team is taking

- deadlines for reaching objectives

- results expected

By keeping this information visible, the activity board should make it easy to read the team's struggles, morale, and successes.

Components of an Activity Board

An activity board is more than just a bulletin board or notice board. It is a guide to team activities. For this reason, it should contain the following elements:

1. The team name, the names of team members, and the name of the team leader, as well as the company's overall policy or vision.

2. Ongoing results from team activities (graphed by month).

3. The theme addressed by the team. Does the theme pertain to current problems?

4. The current situation and its causes (expressed quantitatively wherever possible).

5. Actions to address those causes and effects of specific actions on team results. Annotating graphs can be an effective way to show the relationship between team activities and results.

6. A log of targets achieved, remaining problems, and actions planned to resolve them.

Autonomous maintenance implementation is strengthened when the activity board also lists autonomous maintenance steps (along with goals), a history of team efforts, reviewers' comments, and a list of problems encountered at each step.

Activity Board

Leader _____

Members _____ _____

_____ _____

Vision **Target policy**

1 -------
2 -------
3 -------
4 -------

Improvement theme

Why we chose this theme

1 ----------------
2 ----------------
3 ----------------

Results
Overall equipment effectiveness
Line 3

% target
(comment)

1 2 3 4 5 6 7 8 9 10 11 12

Availability | Performance | Quality

month | month | month

Current situation

1 --------------
2 --------------
3 --------------

Analysis

1 ------------------------
2 ------------------------
3 ------------------------

Effects

1 ------------------------
2 ------------------------
3 ------------------------

target

Actions

Items	Targets	Methods	Schedule (days)

	----		

10 20 30

Standardization

Remaining issues and future plans

- Activity board
- Meetings
- One-point lessons

Using the Activity Board in Meetings

An activity board is very effective in promoting a sense of team-work because it clarifies the current situation, what needs to be done, and how actions need to proceed. Indeed, an activity board helps team members focus meetings on the right issues and steer team activities in the right direction. Recording the team's progress on an activity board also helps prevent lapses and oversights.

In order to spread and deepen knowledge, activity boards are often supplemented by learning aids called *one-point lessons.* Taken together, activity boards, meetings, and one-point lessons are the essential tools of team activities. (Chapter 6 will focus on the development of one-point lessons.)

Managing Meetings

Regular meetings are indispensable for making team activities a way of life. A lively exchange of views in meetings often leads to exceptional improvement ideas.

It is important, then, to create an atmosphere in which everyone feels free to exchange opinions. The leader needs to manage meetings so that certain members don't dominate the discussion or hold the floor for too long. One way to do this is to give everyone an agenda in advance so they can be prepared to discuss the subjects on the schedule.

Meetings should be short and frequent. The leader should determine beforehand the topic and purpose of the meeting, the expected outcomes, problems to be discussed, and actions to be taken by the group.

95

Team Activity Report	Issued	(date)

Team Activity Report

Issued	(date)
Team name	
Theme:	Affiliation
	Team leader	Scribe

	Project work	__/__ from : to :
Present:	Meetings	__/__ from : to :
	Training	__/__ from : to :
Absent:	Total time	(hours) x (people) = hours

No.	Problems and issues	Actions and countermeasures	When	Who

Department head's comments:	Department office comments:	Group leader comments:
signature	signature	signature

TPM office comments:	Next theme:	Next session scheduled:
signature		Date: Time:

Routing

Issuer → Group leader → Department office → Department head → TPM office

Keeping Minutes of Meetings

Since meetings confirm actions that have been taken, minutes of what takes place in meetings provide valuable documentation of the team's progress. As soon as a meeting is over, a record showing the date and time, the attendees and the leader, topics discussed, decisions made, etc., should be circulated to the team's department manager for comments.

Such minutes help managers understand the issues the team is addressing and spot potential difficulties ahead of time. The advice and encouragement they can give in response make these records a powerful tool for supporting and energizing teams.

Rules for Meetings

1. Time is limited, so stick to the point.

2. The team comes first; focus on team goals.

3. Keep track of the main points of the meeting on a board.

4. Speak with an open mind.

5. Listen with an open mind.

6. Don't be too formal.

7. Don't monopolize the discussion.

8. Avoid talking in vague generalities.

9. Accept contrary views.

10. Express your views when they are fresh.

Ten Tips for Leaders

Before the Meeting Begins

1. Be enthusiastic when you announce the meeting so team members will be energized about it and prepared to work on the issues.

2. Advise your manager of plans for the meeting.

3. Write subjects for discussion and decision in big letters on a board or flip chart paper before the meeting. Prepare relevant data and information.

During the Meeting

4. At the start of the meeting, write the scheduled ending time in big letters on the board.

5. Assign specific people to fill the roles of facilitator, scribe, and activity board manager. The scribe should keep minutes.

6. Go over the agenda of subjects for discussion and decision.

7. For each action decided, confirm the target completion date and the person responsible for completion of the action.

8. Set the date for the next meeting. Thank everyone for their hard work.

After the Meeting

9. Post minutes of the meeting on the activity board and update the team's action plan.

10. Report to your manager and get his or her guidance and advice.

Before the meeting

1. Communicate in advance

2. Report to manager

3. Write on blackboard

Theme...

During the meeting

Theme
Decisions

End
at
6:00

4. Indicate time meeting ends

5. Assign roles

6. Go over theme and decisions

7. Confirm assignments and completion dates

8. Set time for next meeting

After the meeting

9. Fill in activity board

10. Get manager's advice

Conclusion

For successful team activities, think of your activity board as a partner that makes work easier. A good activity board makes people want to participate in autonomous maintenance activities. Using the activity board

- helps everyone work together

- gives team members a sense of accomplishment

- shows everyone the progress they are making

And team meetings run smoother because the activity board

- provides new information

- helps focus discussions

- gives people a place to express themselves

CHAPTER SUMMARY

Autonomous maintenance activities draw on the strength of teamwork for skill building and successful problem solving. Team success depends on motivation, ability, and a favorable environment. Teamwork energizes the workplace and shares the satisfaction of achievement.

Understanding the basic elements of problem solving is a key to team motivation. Stay focused on

1. *What* are we going to do? (Theme)

2. *Why* are we going to do it? (Vision)

3. *How far* are we going to go? (Targets)

4. *How* are we going to do it? (Method)

5. *What is the sequence* and timing of actions? (Schedule)

6. *Who* does what? (Roles)

7. *What results* do we expect? (Assessment)

The activity board is a visual guide to the team's activities that makes these elements so clear that anyone can immediately understand them. It is a visual scoreboard for the team's progress toward company, departmental, and team visions and targets.

The activity board keeps everyone's attention on the project, using color and graphs to make information clear to everyone who sees it. It becomes the natural focus for team meetings, as well as a place for feedback and encouragement from management.

Remembering the goal is a key purpose of the activity board. It shows at a glance the team's goals and the specific problems the team is addressing, actions the team is taking, deadlines for reaching objectives, and results expected. Main elements of the board include

1. Team name, members, and leader; company and departmental policy or vision

2. Ongoing results from team activities, graphed by month

3. Theme of activities

4. Current situation and its causes

5. Actions to address those causes

6. A log of targets achieved, remaining problems, and further actions planned

An activity board helps the team focus meetings on the right issues and steer activities in the right direction. Recording the team's progress on an activity board also helps prevent lapses and oversights.

A lively exchange of views in meetings often leads to exceptional improvement ideas. It is important, then, to create an atmosphere in which everyone feels free to exchange opinions. Meetings should be short and frequent. The leader should determine beforehand the topic and purpose of the meeting, the expected outcomes, problems to be discussed, and actions to be taken by the group.

At every team meeting, keep minutes to record the topics discussed and the decisions made. Circulate these to the department manager for comments.

Following certain guidelines will keep meetings productive. As a group, speak and listen with open minds, focusing on team goals. Leaders can set the stage for effective meetings by communicating and preparing in advance, reviewing roles and agenda at the start, confirming action items, and posting and communicating the minutes afterward. The activity board is a useful focal point during and between meetings.

One-Point Lessons

CHAPTER 6

CHAPTER OVERVIEW

One-Point Lessons

- • What Is a One-Point Lesson?

- • One-Point Lessons as a Cascading Training Tool

- • Purposes and Philosophy

- • Types of One-Point Lessons

- • The One-Point Lesson Form

- • Tips for Writing One-Point Lessons

- • Presenting the One-Point Lesson

- • TPM One-Point Lesson Examples

- • Chapter Summary

What Is a One-Point Lesson?

A one-point lesson is a 5- to 10-minute self-study lesson drawn up by team members and covering a single aspect of equipment or machine structure, functioning, or method of inspection.

105

> "It is frequently difficult to devote much continuous time to training. At the same time, much that is learned is forgotten because there is no opportunity to practice it. Learning in short periods of time during daily meetings or production activities is very effective in overcoming these problems. One-point lessons are a training tool that can be effectively incorporated into ongoing autonomous maintenance activities."
>
> —from *The TPM Dictionary*
> (Japan Institute of Plant Maintenance)

One-Point Lessons as a Cascading Training Tool

Team leaders and members who have special training or knowledge about equipment need a way to share their knowledge with their teammates. Rather than merely repeating what they have been taught, they need to put it in a form that suits their particular workplace. This is where one-point lessons are useful.

One-point lessons are an extremely effective teaching tool because they are short and focused on a topic team members need to know about. They offer a simple vehicle for going over the material until everyone has mastered it. The act of teaching develops leadership skills on the team.

At the same time, it is not enough to simply pass knowledge on to team members. It is essential that the person who gives the instruction, the team leader, or the team as a group follow up to ensure that the knowledge is put into practice every day.

Purposes and Philosophy

One-point lessons have three purposes:

1. To help sharpen equipment-related knowledge and skills and communicate information about specific problems and improvement

2. To share important information easily when it is needed

3. To improve the performance of the entire team

The basic philosophy behind one-point lessons is simple:

1. Develop and research the lessons yourselves.

2. Make up your own lesson sheets.

3. Explain them to all team members.

4. Discuss them openly on the shop floor.

5. Improve them.

109

Types of One-Point Lessons

Depending on their purpose, one-point lesson sheets fall into one of three categories:

1. Basic knowledge lessons

2. Examples of problems

3. Examples of improvements

It is important to keep these categories distinct.

Basic Knowledge Lessons

These are training tools designed to fill in knowledge gaps and ensure that team members have the knowledge they need for daily production and for TPM activities. These lessons may focus on equipment subsystems, safety points, or basic operating information.

Examples of Problems

Based on problems that have actually occurred, these lessons are designed to communicate knowledge or skills to help operators prevent similar problems from happening in the future.

Examples of Improvements

To ensure that successful improvement ideas are used widely, these lessons present what needs to be done to prevent or correct equipment abnormalities by describing the approaches, actions, and results of specific improvement projects.

```
┌─────────────────────────────────────────────────────────────┐
│ ┌──────────┐                          ☐ Basic knowledge      │
│ │ One-Point │                          ☐ Problem example      │
│ │  Lesson  │                           ☐ Improvement example  │
│ └──────────┘                                                  │
│ Theme: _____ │
│ _____ │
│                                                               │
│                                                               │
│                                                               │
│                                                               │
│                                                               │
│                                                               │
│                                                               │
│                                                               │
│                                                               │
│                                                               │
│                                                               │
│                                                               │
│                                                               │
│ Date developed_____   Developer _____  │
│ Revised _____   Dept./Team _____   │
│ TPM No. _____                                     │
└─────────────────────────────────────────────────────────────┘
```

The One-Point Lesson Form

The form shown here is a generic form that could be used for any of the types of one-point lessons.

You can customize your own one-point lesson form to help people organize the information they need to share. For instance, a form for problem examples might include separate areas to record a description of the problem, the cause of the problem, the action taken, and preventive steps. A form for examples of improvements might be divided into areas for an explanation of current problems and a description of improvements and results.

If you will be circulating the lesson to other areas of the company, you may want to include a space for tracking where it is used or posted.

Tips for Writing One-Point Lessons

To make sure everyone shares a common understanding of the lessons, keep in mind the following points when making up the lessons:

113

- Choose a theme based on a common problem the workplace is currently facing.

- Don't just describe the problem in words. Feel free to use drawings, photographs, or cartoons so everyone can grasp what is important.

- If you model your forms on those from other teams or other workplaces, be sure to adapt them to your particular needs.

- For examples of problems, make up the lesson immediately after the problem has occurred and then teach it while the issue is fresh in everyone's minds.

Presenting the One-Point Lesson

Here is how a one-point lesson should be presented:

1. Present the theme and explain the motivating reason for writing up the lesson.

2. As you go through the lesson, ask questions of the group; try to get team members to examine their own knowledge and behavior relating to the theme.

3. Don't rely on only the one-point lesson sheet. Demonstrate the lesson using the actual objects or parts involved wherever possible.

4. Ask people questions after the presentation and follow up to make sure that everyone has understood.

5. Repeat the lesson several times if necessary, until you are sure that it has been linked to action. For each presentation, keep track of the date and the people attending by noting them on the form.

114

TPM One-Point Lesson Examples

TPM One-Point Lesson (improvement example)

Theme:	Improvement of Unreadable Oil Level Gauge on Brake

Current Situation

The oil gauges on a conveyor brake and a gear motor

- can't be checked because they are hidden behind pulleys

- are discolored and cloudy

Spares are difficult to get and hard to install, so the gauges aren't checked and lubrication is left unmanaged.

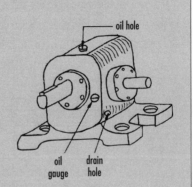

Description of Improvement

We attached a commercially available oil gauge to a short pipe inserted in the drain hole. This made it easy to check the oil level.

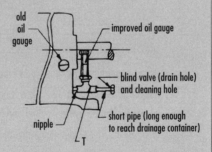

Key Points to Note

1. Don't attach an L-shaped oil gauge directly to the drain hole; sludge will accumulate in the bend and eventually clog the level.

   - Instead, attach the oil gauge to a drain pipe inserted in the drain hole. This way, any sludge can be flushed out.

2. Permanently mark the new oil gauge with the same level as the old gauge.

TPM One-Point Lesson Examples (continued)

TPM One-Point Lesson (basic knowledge)

Theme:	Is the hydraulic unit working normally? (Part 1)

If you can't tell whether a hydraulic unit is functioning normally, check hydraulic fluid temperature and machine cycle time the day after a day off (on a Monday, for example) and make up a graph like the one below.

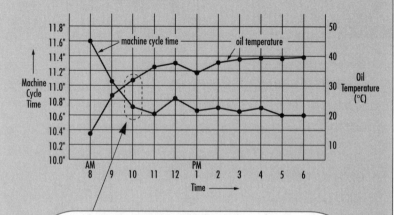

On some machines, machine cycle time may not stabilize until the oil temperature rises over 30°C. This is because sludge from machining has contaminated hydraulic fluid and sticks to pipes and hydraulic mechanisms. This sludge generally melts at around 30°C.

If your hydraulic unit shows machine cycle time stabilizing at about 30°C as in the graph, do the following:

    1. Thoroughly clean out the hydraulic fluid reservoir.

    2. Flush out pipes.

The next thing to look at is prevention.

    • Take steps to prevent contamination of the hydraulic unit.

TPM One-Point Lesson Examples (continued)

TPM One-Point Lesson (basic knowledge)	
Theme:	Is the hydraulic unit working normally? (Part 2)

If machine cycles remain slow at low temperatures even after cleaning out the hydraulic unit as described in Part 1, you need to move to the next level of countermeasures. But here are some questions to ask before you do:

1. What type of hydraulic parts are used in the unit?

2. Does the unit conform to the hydraulic routing diagram? (Check pipe diameters, for example.)

Older flow control valves (20-50 l/min. or less) are typically not equipped with temperature compensators. As a result, changes in hydraulic fluid viscosity due to temperature fluctuations may cause fluidity to vary by 10 to 20 percent.

In environments where such cycle time fluctuations pose a problem, try installing a flow controller equipped with a temperature compensator mechanism.

TPM One-Point Lesson Examples (continued)

TPM One-Point Lesson (example of problem)

Theme:	Broken key on hobbing machine hydraulic pump

Phenomenon

- Hydraulic pressure (set at 30 kgf/cm²) on workpiece clamping cylinder dropped and did not come back up.

Cause

- The semicylindrical key on the hydraulic pump broke and force from the motor was not transmitted.
- This was because the centers of the hydraulic pump and the motor were not aligned.
- The misalignment was caused by a loose clamping bolt on the hydraulic pump.

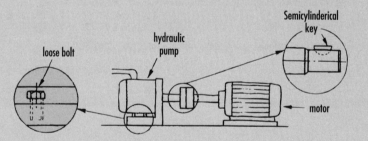

Actions

1. Corrected misalignment of pump and motor (0.5 mm→0.1 mm).
2. Changed semicylindrical key.
3. Tightened clamping bolt on hydraulic pump.
4. Reset hydraulic pressure (relief pressure) to 30 kgf/cm².

Prevention

1. Drew registration marks on hydraulic pump and motor clamping bolts→check once every six months.
2. Check pressure gauge at the start of each shift.

Downtime loss

- approximately 8 hours

CHAPTER SUMMARY

In addition to activity boards, one-point lessons support team activities and learning on specific topics. A one-point lesson is a 5- to 10-minute self-study lesson drawn up by team members to cover a single aspect of equipment or machine structure, functioning, or method of inspection. Learning in short periods of time during daily meetings or production activities is very effective in relating the lesson to actual practice. One-point lessons are an effective training tool for ongoing autonomous maintenance activities.

One-point lessons are extremely effective because they are short and focused on a topic team members need to know about. They offer a simple vehicle for going over the material until everyone has mastered it. The act of teaching develops leadership skills on the team. The team also must follow up to ensure that the knowledge is put into daily practice.

One-point lessons have three purposes:

1. To help sharpen equipment-related knowledge and skills and communicate information about specific problems and improvement

2. To share important information easily when it is needed

3. To improve the performance of the entire team

Depending on their purpose, one-point lesson sheets fall into one of three categories:

Basic knowledge lessons: Fill in knowledge gaps and ensure people have knowledge needed for daily production.

Examples of problems: Communicate knowledge or skills needed to prevent problems.

Examples of improvements: Tell how to prevent or correct equipment abnormalities by detailing the approaches, actions, and results of specific improvement projects.

Keep in mind the following points when making up a one-point lesson:

- Choose a theme based on a common problem the workplace is currently facing.

- In addition to words, use drawings, photographs, or cartoons so everyone can grasp what is important.

- Adapt model forms or lessons to your particular needs.

- For examples of problems, make up the lesson immediately after the problem has occurred and teach it while the issue is fresh.

Here is how a one-point lesson should be presented:

1. Present the theme and explain the motivating reason for writing up the lesson.

2. As you go through the lesson, ask questions of the group; try to get team members to examine their own knowledge and behavior relating to the theme.

3. Don't rely on only the one-point lesson sheet. Demonstrate the lesson using the actual objects or parts involved wherever possible.

4. Ask people questions after the presentation and follow up to make sure that everyone has understood.

5. Repeat the lesson several times if necessary, until you are sure that it has been linked to action. For each presentation, keep track of the date and the people attending by noting them on the form.

Additional Resources

Books and Learning Packages

Japan Institute of Plant Maintenance, ed., *TPM for Every Operator* (Productivity Press, 1996). Explains TPM activities carried out by operators, including autonomous maintenance, focused improvement, and safety activities. Introduces the six big losses in easy-to-grasp terms.

Productivity Press Development Team, *5S for Operators* and *5S for Operators Learning Package* (Productivity Press, 1996). Introduces operators to the 5S system for workplace organization, cleaning, and standardization—the foundation for TPM and other advanced improvement approaches. Learning Package includes leader's discussion guide, overheads, color slides of examples, and resource books.

Productivity Press Development Team, *TPM for Every Operator Learning Package* (Productivity Press, 1997). Provides a framework for group learning about TPM and autonomous maintenance basics, as well as application exercises to see how TPM concepts might apply in the learners' work areas. Includes leader's discussion guide and overheads, study guides for participants, and exercise worksheets.

Productivity Press Development Team, *TPM for Supervisors* (Productivity Press, 1996). Presents the basic methodology of TPM, with a focus on operator activities to reduce equipment-related losses and maximize equipment effectiveness.

K. Shirose, ed., *TPM Team Guide* (Productivity Press, 1996). A shopfloor series book that teaches how to lead TPM team activities in the workplace and develop effective presentations of project results.

K. Shirose, *TPM for Workshop Leaders* (Productivity Press, 1992). Describes the hands-on leadership issues of TPM implementation for shopfloor TPM group leaders, with case studies and practical examples to help support autonomous maintenance activities.

Training and Consulting

The Productivity Consulting Group delivers hands-on public events and customized in-house training and consulting to support autonomous maintenance and TPM implementation. Telephone: 1-800-966-5423.

About the Editor

The Japan Institute of Plant Maintenance is a nonprofit research, consulting, and educational organization that helps companies increase organizational efficiency and profitability through improved maintenance of manufacturing equipment, processes, and facilities. The JIPM is the sponsoring organization for the PM Prize, awarded annually to recognize excellence in companywide maintenance systems. Based in Japan, JIPM is the innovator of methodologies that have been implemented around the world. Productivity Press is pleased to be the publisher of the English editions of many groundbreaking JIPM publications.

ABOUT THE SHOPFLOOR SERIES

Put powerful and proven improvement tools in the hands of your entire workforce!

Progressive shopfloor improvement techniques are imperative for manufacturers who want to stay competitive and to achieve world class excellence. And it's the comprehensive education of all shopfloor workers that ensures full participation and success when implementing new programs. The Shopfloor Series books make practical information accessible to everyone by presenting major concepts and tools in simple, clear language and at a reading level that has been adjusted for operators by skilled instructional designers. One main idea is presented every two to four pages so that the book can be picked up and put down easily. Each chapter begins with an overview and ends with a summary section. Helpful illustrations are used throughout.

Books currently in the Shopfloor Series include:

5S FOR OPERATORS
5 Pillars of the Visual Workplace
The Productivity Press Development Team
ISBN 1-56327-123-0 / 133 pages
Order 5SOP-BK / $25.00

QUICK CHANGEOVER FOR OPERATORS
The SMED System
The Productivity Press Development Team
ISBN 1-56327-125-7 / 93 pages
Order QCOOP-BK / $25.00

MISTAKE-PROOFING FOR OPERATORS
The Productivity Press Development Team
ISBN 1-56327-127-3 / 93 pages
Order ZQCOP-BK / $25.00

JUST-IN-TIME FOR OPERATORS
The Productivity Press Development Team
ISBN 1-56327-134-6 / 96 pages
Order JITOP-BK / $25.00

TPM FOR EVERY OPERATOR
The Japan Institute of Plant Maintenance
ISBN 1-56327-080-3 / 136 pages
Order TPMEO-BK / $25.00

TPM FOR SUPERVISORS
The Productivity Press Development Team
ISBN 1-56327-161-3 / 96 pages
Order TPMSUP-BK / $25.00

TPM TEAM GUIDE
Kunio Shirose
ISBN 1-56327-079-X / 175 pages
Order TGUIDE-BK / $25.00

AUTONOMOUS MAINTENANCE
The Japan Institute of Plant Maintenance
ISBN 1-56327-082-x / 138 pages
Order AUTOMOP-BK / $25.00

FOCUSED EQUIPMENT IMPROVEMENT
FOR TPM TEAMS
The Japan Institute of Plant Maintenance
ISBN 1-56327-081-1 / 144 pages
Order FEIOP-BK / $25.00

OEE FOR OPERATORS
The Productivity Press Development Team
ISBN 1-56327-221-0 / 96 pages
Order OEEOP-BK / $25.00

CELLULAR MANUFACTURING
One-Piece Flow for Workteams
The Productivity Press Development Team
ISBN 1-56327-213-X / 96 pages
Order CELL-BK / $25.00

KANBAN FOR THE SHOPFLOOR
The Productivity Press Development Team
ISBN 1-56327-269-5 / 120 pages
Order KANOP-BK / $25.00

KAIZEN FOR THE SHOPFLOOR
The Productivity Press Development Team
ISBN 1-56327-272-5 / 112 pages
Order KAIZOP-BK / $25.00

PULL PRODUCTION FOR THE SHOPFLOOR
The Productivity Press Development Team
ISBN 1-56327-274-1 / 122 pages
Order PULLOP-BK / $25.00

STANDARD WORK FOR THE SHOPFLOOR
The Productivity Press Development Team
ISBN 1-56327-273-3 / 112 pages
Order STANOP-BK / $25.00

www.productivitypress.com

OTHER BOOKS FROM PRODUCTIVITY PRESS

Productivity Press publishes books that empower individuals and companies to achieve excellence in quality, productivity, and the creative involvement of all employees. Through steadfast efforts to support the vision and strategy of continuous improvement, Productivity Press delivers today's leading-edge tools and techniques gathered directly from industry leaders around the world. Call toll-free 1-800-394-6868 for our free catalog.

Implementing TPM
The North American Experience
Charles J. Robinson and Andrew P. Ginder

This book offers a modified approach to TPM planning and deployment that builds on the 12-step process advocated by the Japan Institute of Plant Maintenance. More than just an implementation guide, it's actually a testimonial of proven TPM success in North American companies through the adoption of "best in class" manufacturing practices. Of special interest are chapters on implementing TPM in union environments, integrating benchmarking practices to support TPM, and a requirements checklist for computerized maintenance management systems.

ISBN 1-56327-087-0 / 224 pages / $45.00 / Order IMPTPM-B280

The Visual Factory
Building Participation Through Shared Information
Michel Greif

If you're aware of the tremendous improvements achieved in productivity and quality as a result of employee involvement, then you'll appreciate the great value of creating a visual factory. This book shows how visual management can make the factory a place where workers and supervisors freely communicate and take improvement action. It details how to develop meeting and communication areas, communicate work standards and instructions, use visual production controls such as kanban, and make goals and progress visible. Includes more than 200 diagrams and photos.

ISBN 0-915299-67-4 / 305 pages / $55.00 / Order VFAC-B280

Poka-Yoke
Improving Product Quality by Preventing Defects
Nikkan Kogyo Shimbun Ltd. and Factory Magazine (ed.)

If your goal is 100 percent zero defects, here is the book for you—a completely illustrated guide to poka-yoke (mistake-proofing) for supervisors and shop-floor workers. Many poka-yoke devices come from line workers and are implemented with the help of engineering staff. The result is better product quality—and greater participation by workers in efforts to improve your processes, your products, and your company as a whole.

ISBN 0-915299-31-3 / 295 pages / $65.00 / Order IPOKA-B280

www.productivitypress.com

A Revolution in Manufacturing
The SMED System
Shigeo Shingo

The heart of JIT is quick changeover methods. Dr. Shingo, inventor of the Single-Minute Exchange of Die (SMED) system for Toyota, shows you how to reduce your changeovers by an average of 98 percent! By applying Shingo's techniques, you'll see rapid improvements (lead time reduced from weeks to days, lower inventory and warehousing costs) that will improve quality, productivity, and profits.

ISBN 0-915299-03-8 / 383 pages / $75.00 / Order SMED-B280

Toyota Production System
Beyond Large-Scale Production
Taiichi Ohno

Here's the first information ever published in Japan on the Toyota production system (known as Just-In-Time manufacturing). Here Ohno, who created JIT for Toyota, reveals the origins, daring innovations, and ceaseless evolution of the Toyota system into a full management system. You'll learn how to manage JIT from the man who invented it, and to create a winning JIT environment in your own manufacturing operation.

ISBN 0-915299-14-3 / 163 pages / $45.00 / Order OTPS-B280

TPM in Process Industries
Tokutaro Suzuki (ed.)

Process industries have a particularly urgent need for collaborative equipment management systems like TPM that can absolutely guarantee safe, stable operation. In TPM in Process Industries, top consultants from JIPM (Japan Institute of Plant Maintenance) document approaches to implementing TPM in process industries. They focus on the process environment and equipment issues such as process loss structure and calculation, autonomous maintenance, equipment and process improvement, and quality maintenance. Must reading for any manager in the process industry.

ISBN 1-56327-036-6 / 400 pages / $85.00 / Order TPMPI-B280

Uptime
Strategies for Excellence in Maintenance Management
John Dixon Campbell

Campbell outlines a blueprint for a world class maintenance program by examining, piece by piece, its essential elements—leadership (strategy and management), control (data management, measures, tactics, planning and scheduling), continuous improvement (RCM and TPM), and quantum leaps (process reengineering). He explains each element in detail, using simple language and practical examples from a wide range of industries. This book is for every manager who needs to see the "big picture" of maintenance management. In addition to maintenance, engineering, and manufacturing managers, all business managers will benefit from this comprehensive and realistic approach to improving asset performance.

ISBN 1-56327-053-6 / 204 pages / $35.00 / Order UP-B280

www.productivitypress.com